MW01045139

Food Found All Around

by Janine Scott

Content and Reading Adviser: Joan Stewart
Educational Consultant/Literacy Specialist
New York Public Schools

Compass Point Books
Minneapolis, Minnesota

Compass Point Books
3722 West 50th Street, #115
Minneapolis, MN 55410

Visit Compass Point Books on the Internet at *www.compasspointbooks.com*
or e-mail your request to *custserv@compasspointbooks.com*

Photographs ©:
PhotoDisc, cover (grass, rabbit); Two Coyote Studios/Mary Walker Foley, cover (worms); Corel, cover (coyote); PhotoDisc, 4, 5; Two Coyote Studios/Mary Walker Foley, 6, 7 (bird); Jennifer Staplerhands, 7 (cat); Corbis, 8; Two Coyote Studios/Mary Walker Foley, 9, 10; PhotoDisc, 11, 12, 13; Two Coyote Studios/Mary Walker Foley, 14, 15; PhotoDisc, 16, 17; Visuals Unlimited/David B. Fleetham, 19; Two Coyote Studios/Mary Walker Foley, 20.

Project Manager: Rebecca Weber McEwen
Editor: Alison Auch
Photo Researcher: Jennifer Waters
Photo Selectors: Rebecca Weber McEwen and Jennifer Waters
Designer: Mary Walker Foley

Library of Congress Cataloging-in-Publication Data

Scott, Janine.
Food found all around / by Janine Scott.
p. cm. -- (Spyglass books)
Includes bibliographical references (p.).
ISBN 0-7565-0234-9 (hardcover)
1. Food chains (Ecology)--Juvenile literature. [1. Food chains (Ecology) 2. Ecology.] I. Title. II. Series.
QH541.14 .S384 2002
577'.16--dc21

2001007334

Printed in the United States of America.

Contents

Links in the Chain

Every plant and animal on the land, in the water, and under the ground plays an important role in a food chain.

An eagle eats meat.

A cow eats grass.

Some food chains are simple.
Leaves grow on a tree.
A caterpillar eats the leaves.
Then a bird eats
the caterpillar.
Then a cat eats the bird.

Leaves

Caterpillar

A Food Chain

Bird

Cat

The First Link

The first link in a food chain
is the green plant.
Green plants make
their own food.
Then the plant becomes food
for plant-eating animals.

Plant leaves take in energy from sunshine. Their roots take in water.

Tree leaves are food for many different kinds of animals.

Plant Eaters

The second link in a food chain is the plant eater, or ***herbivore***.
Plant eaters cannot make their own food.
They must eat plants to ***survive***.

A giraffe eats leaves.

A sheep eats grass.

A chipmunk
eats seeds.

Meat Eaters

Meat eaters, or ***carnivores***, are the third link in a food chain. They eat other animals. Animals that eat both animals and plants are called ***omnivores***.

Cats have sharp teeth and claws for eating meat.

Tiger

Decomposers

The last link in a food chain is the ***decomposer***. Decomposers eat dead plants and animals.

They leave behind things in the soil that help plants grow.

Then the food chain cycle begins again.

Did You Know?

Earthworms are decomposers. After they eat dead plants, they leave behind rich soil that helps new plants grow.

On the Land

Each area on Earth has
its own special food chains.
In the African grasslands,
zebras feed on grass.
Zebras are then eaten by lions
or hyenas.

Lion

Did You Know?

Tiny animals are part of food chains, too. In a garden, plants are eaten by aphids. Aphids are eaten by ladybugs.

Zebras

Under the Water

Food chains can also be found in oceans, rivers, streams, and ponds.

Blue whales eat tiny animals called “krill,” but the food chain ends there. No other animals, except humans, hunt the whales.

This blue whale has come up for air after diving to find krill to eat.

Backyard Food Chain

1. Think of a simple food chain that might be in your yard, or in a nearby park.

2. Draw a box for each thing. Draw arrows going from one box to the next.

3. Write the names of the things you thought of in the boxes.

An example might be:

Glossary

carnivore—an animal that eats other animals

decomposer—a living thing that eats dead plants and animals, and then breaks them down into tiny parts that make food for new plants to grow

herbivore—an animal that eats plants

omnivore—an animal that eats both plants and animals

survive—to stay safe and alive

Learn More

Books

Butler, Daphne. *Gathering Food*. Austin, Tex.: Raintree Steck-Vaughn Publishers, 1995.

Hickman, Pamela. *Hungry Animals: My First Look at a Food Chain.* Illustrated by Heather Collins. Toronto, Ontario: Kids Can Press Ltd., 1997.

Jenkins, Steve. *What Do You Do When Something Wants to Eat You?* Boston: Houghton Mifflin, 1997.

Web Site

Brain Pop

www.brainpop.com/science/seeall.weml (click on "food chain")

Index

GR: I
Word Count: 220

From Janine Scott

I live in New Zealand, and have two daughters. They love to read fact books that are full of fun facts and features. I hope you do, too!